I0134958

THIS BOOK COULD NOT HAVE BEEN COMPLETED WITHOUT THE AIDE OF ZOO KNOXVILLE AND ITS STAFF. SPECIAL THANKS IN PARTICULAR TO (IN ALPHABETICAL ORDER BY FIRST NAME) AMY FLEW, MICHAEL OGLE, RACHEL VARNELL, AND SARAH GLASS.

PAPERBACK ISBN: 978-0-9967189-9-8
LCCN: 2017918421

PHOTO CREDITS

ALL IMAGES NOT OTHERWISE GIVEN CREDIT TO WERE TAKEN BY DEBORAH JOHNSON

FOR MORE BOOKS BY N.A. CAULDRON, PLEASE VISIT HTTP://NACAULDRON.COM

## Other Works by N.A. Cauldron

*The Cupolian Series*

*Fishing for Turkey*

If you enjoyed this book, please leave a review.

http://nacauldron.com/

Dear Reader,

Please understand that not only is it impossible to include a complete collection of different **species feces** within this book, it is also impossible to describe all poop in detail. Therefore, I leave it to you, dear Reader, to utilize the fine tools of the internet and libraries to further your study of this subject, where you will find a plethora of various poop, both in still frame and in video form. At the end of this book, you will find links to other websites you can use to further your exploration into this topic. Please understand that links expire and websites change, so I do apologize if any of the links I provide have done just that. You will also find a glossary. When a glossary term is used for the first time, it is in bold.

I have placed several questions throughout the text. Please, try to answer them before moving on. Many of them are like a scavenger hunt within the poop (yes, you read that correctly) and will aide you in your learning.

I normally prefer real books, but I highly recommend the digital format for this one. My ebooks are normally free with the purchase of a physical copy, so please take advantage of that. The reason for this is that I include many outside links to videos and such you may want to visit while reading this book. If you have the paper version, you will have to type the links in manually. With the e-version, all you have to do is click.

I can only hope that this book inspires you enough to learn more of the world of poop. Enjoy!

Sincerely,
Ms Cauldron

## THANK YOU

I want to give a big thank you to Zoo Knoxville and the people that work there, Amy Flew, Michael Ogle, Rachel Varnell, and Sarah Glass. Without them, this book would not have been possible.

## TABLE OF CONTENTS

# THE PURPOSE OF STUDYING POOP

In this book, we will not only see the varying types of poop out there, but we will also learn why it looks the way it does.

Let's begin with the most common poop to study, **scat.**

(Scat of a red fox)

Scat is the poop of wild animals, and it comes in a variety of shapes, colors, and yes, flavors. According to the Merriam Webster Dictionary, the first known use of scat as the meaning of animal excrement was in 1927. Prior to that, it was solely used to shoo away kitties. The first time[1] someone told a cat to "Scat!" was in 1838.

---

[1] http://www.merriam-webster.com/dictionary/scat

Taking scat apart and looking at it can tell you what the animals who produced it have eaten, and therefore give you a better understanding of where the animal has been, how healthy the population is, and if any diseases are plaguing them. It's a very useful tool, especially for those who take care of our wildlife. If you enjoy this book, and think you might like to work in a field where studying animal poop is part of your job, I encourage you to look into the different careers in wildlife management. The websites listed at the end of this book can help you with that.

Do you have some scat in your backyard? You probably do and don't even know it! Want to learn more about scat? Jeff the Nature Guy has an excellent video on it. This video is best viewed after reading this book, so its link has been placed at the end.

## BIRDS

Everyone's seen bird poop. It lands on cars, sidewalks, and even people's hair! But have you ever wondered why it's two different colors? Why doesn't it look like your poop? Let's find out.

See that picture? That's poo from my own chickens. I've raised countless chickens, ducks, and turkeys, and other than the size of it, all their poo looks the same. But what about an ostrich?

When you think about an ostrich, you probably think that since it's a big animal, it should **excrete** big, brown poo, you know, like a cow or something. But it doesn't. Ostrich poop is just like any other bird, only much, much bigger.

All birds have white stuff, and brown/black stuff in their poop. But why?

When a person, such as yourself, goes potty, your **kidneys** pull out the wastes from your blood, mix them with water, and then let it all go in the form of **urine**, that yellow stuff you probably call pee. Birds do the same thing, only without so much water. Their pee is actually uric acid (ow!) and therefore white. It's called **urate**. That dark stuff you see in the white, that's their actual poo.

But, and this is the really fun part, a bird doesn't have a bunch of plumbing down there like we do. It all comes out of the same place. That place is called a **cloaca**. And when they lay an egg? Yup, same place. Just think about that next time you order fried eggs for breakfast!

# CYRANO GREEN ARACARI

The green aracari is a toucan from South America. Its poop looks very different from a chicken's.

Can you guess why? The reason is diet. A toucan eats different stuff than the chicken does. While a chicken will eat fruits, vegetables, and **insects**, it is unable to fly and can only eat what it can find on the ground. The toucan lives in the trees and is able to have a more fruit and nut based diet than the chicken. Can you tell what this toucan was recently fed by looking at its poop?

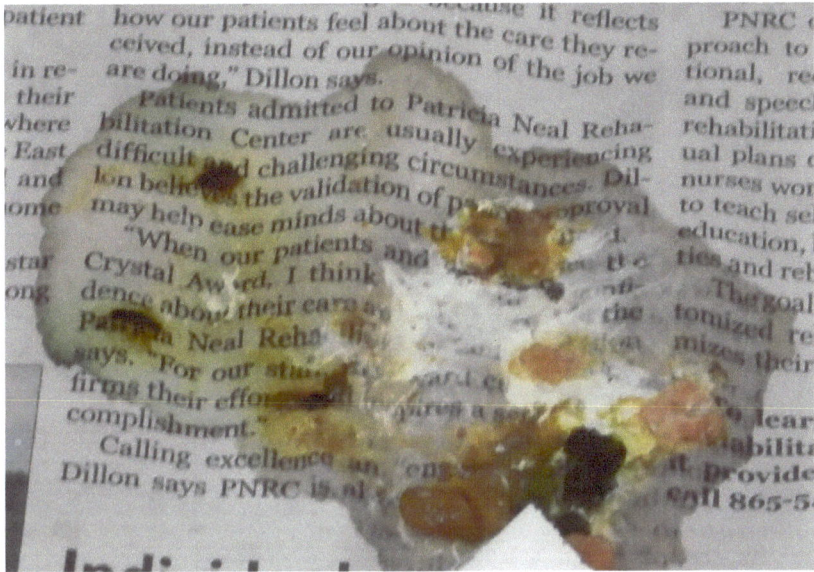

See the different colors and shapes of the seeds? Those are the parts that didn't **digest** all the way. That's because seed **hulls** don't digest very well. It's things like this that help us discover what an animal's diet is.

Some people keep these birds as pets. To see a pet aracari playing with a Golden Retriever, watch this video here:
https://youtu.be/ZU3xc6q2QAA

**BARN OWL**

Owls, along with storks, hornbills, and many other species of birds, pass what's called a pellet. Owls eat their food whole and are unable to digest certain items, such as the hair and bones. These items are **regurgitated**, that's a real fancy word for thrown up, in the form of a pellet. Pellets dry out pretty quickly, and a really fun project is to dissect them to see what the owl has recently eaten. You can sometimes find whole skeletons inside one pellet. The end of this book tells where you can search to get your own pellet.

Here's a video of a barred owl regurgitating a pellet:

https://youtu.be/PTLaxqmcYeY
Owls don't just regurgitate pellets. They poop like all the other birds too, except it's mostly white. Can you think why?

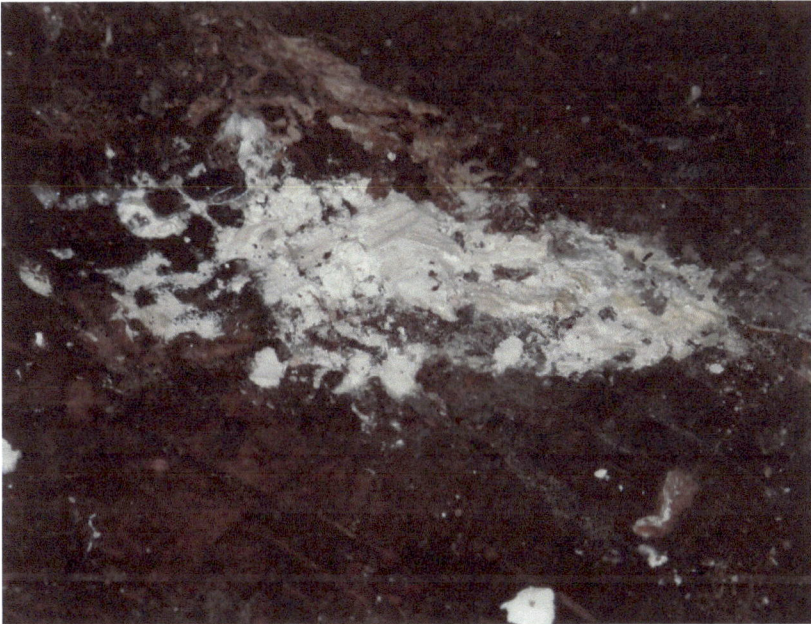

If you'll remember, the black part of bird poop is the solid stuff, like your poop, and the white part is their pee, or urates. Since an owl regurgitates so much of its solid matter, it mainly just excretes urates.

# CRESTED SCREAMER

Photo by Trisha Shears

Obviously, the crested screamer likes to scream. You would too if your poop was purple. What?! The poop of the crested screamer can be purple? Yes, crested screamer poop is often times purple.[2]

For a video of a crested screamer "screaming", visit
https://youtu.be/SeaByXgH7nU

---

[2] Rachel Varnell (Bird Keeper) and Amy Flew (Mammal Curator), interviewed by Deborah Johnson, Knoxville, TN, April 2016

Crested screamer poop is tapered at one end, giving it the appearance of a long chocolate kiss. Mmmm, yummy. This is also what peacock poop looks like.

**CRANE**

Photo Courtesy of A. Basit of Animal Wildlife

Cranes are about the size of a peacock or the crescent screamer, but their poop looks more like regular bird poop than chocolate kisses.

## BALI MYNAH

Originally from Indonesia, the Bali mynah is an **endangered** species. They can't fly very far, and even though it's illegal, a lot of people keep them as pets. They make their nests in holes in the trunks of trees and eat mainly fruit and insects. The mom and the dad share the chores of constructing their nest, sitting on their eggs, and raising their chicks.[3]

The Bali mynah, golden taffeta weaver, and superb starlings all have **toxic** poop, that means poisonous (Who wouldn't want toxic poop?). Their poop isn't always toxic. It's just when they're stressed that

---

[3] Zoo Knoxville

**toxins** are emitted with their poop. But since that means their poop could be toxic at any time, Zoo Knoxville must keep them constantly **quarantined,** or separate from the other animals. While the toxins are obviously not harmful to them personally, this poop can be a great weapon to use against their predators[4].

Unlike what our imaginations would lead us to believe, toxic poop does not

---

[4] Rachel Varnell (Bird Keeper) and Amy Flew (Mammal Curator), interviewed by Deborah Johnson, Knoxville, TN, April 2016

glow green or purple. (Although, that would be pretty awesome!) Toxic poop looks just like normal poop, and that is why it is so dangerous, and so important to treat in a safe manner.

## REPTILES

Reptile poop is very similar to bird poop. It is white and black just like birds, and the white part is also called urate. This is because reptiles use a cloaca like birds do. Remember what a cloaca is?

**COMMON LIZARD**

See the white spot on the rock? That is a urate deposit from a common lizard. The hot sun and dry air turn it into a flat powder, making it look almost like someone spray painted the rock.

## GUATAMALAN BEARDED LIZARD

This is the poop from a Guatemalan bearded lizard. The urate is easily seen next to its regular dark poop. Do you see the similarities between this and the chicken poop near the beginning of the book?

What about how different it is? See how the dark parts have hair in them? That's because the bearded lizard eats mice and other animals with fur. The hair doesn't digest completely and comes out in their poop.

# KOMODO DRAGON

Photo by Michael Ogle

Can you tell where the urates are in this picture? That's correct. It's that white stuff to the right of the logs of poop.

Komodo dragons know all about poop. In fact, after they kill their prey, they will sling the intestines around until all the poop comes out. Then, they will eat the empty intestines. Some dragons will even roll around in their prey's poop in order to smell gross. That way bigger animals won't try to eat them as much.[5]

---

[5] Fuzzy Wuzzy Animpals, dir. "Komodo Dragons and Poop Animal Fact" *YouTube*. March 8, 2013. Web. November 7, 2017.

**SNAKE**

    Snakes are just like lizards when it comes to their poop. Their urate and poop is mixed into a single dropping.
    Here's a video of a huge snake pooping.
    https://youtu.be/LGm3tKOwWX4

# INSECTS, ARACHNIDS, AND OTHER CREEPY CRAWLIES

## HISSING COCKROACH

As you can see, hissing cockroaches poop wherever they happen to be. All that black stuff? Yup that's poop.

Some people keep hissing cockroaches as pets, letting them walk all over their arms and everything. Their feet aren't as prickly as they look, so it doesn't feel as creepy weird as you would expect. Hey, at least they don't bark at the neighbors.

# TARANTULA

See that white spot? That's tarantula poop. You can even see the spider's leg in the lower right corner of the picture.

Tarantulas, like most spiders and reptiles, do not eat every day. When they do eat, they are very clean. They have a set apart area of their habitat that is their "dumping ground". Once finished with their food, they roll what they don't want up with dirt or whatever material they have on hand, and place the neat little package with all the other garbage.[6]

---

[6] Sarah Glass (Curator), interviewed by Deborah Johnson, Knoxville, TN, April 2016

**CRICKET**

This is cricket poop. I'll let you in on a little secret. It smelled like oranges. Cool! Right? That was because they were mainly fed a **citrus** diet. These crickets were being raised to feed other animals at the zoo, like that tarantula you just read about.[7]

Are you wondering if grasshoppers poop the same way? Follow this link to find out.
https://youtu.be/JsdEa_3L8GQ

---

[7] Sarah Glass (Curator), interviewed by Deborah Johnson, Knoxville, TN, April 2016

## COMMON HOUSE SPIDER

    Unlike the tarantula, the common house spider has tiny dollops of brown poop, as seen here. These spiders are very useful for eating gnats, flies, and other bothersome insects within your house. I keep several "pet" house spiders, cleaning their poop up as needed, in order to help with the insect population of my home.

    There are lots of videos out there of spiders pooping, so please, go find them!

# HORN WORM

Photo by Jay and Melissa Malouin

The horn worm comes as the tobacco horn worm, or the tomato horn worm.[8] I have personally had to deal with these unpleasant creatures on my own tomato plants. Due to their enormous size, they can destroy a tomato plant very quickly indeed! However, do not be alarmed. Although large and scary looking, they are harmless to humans.[9]

---

[8] *Featured Creatures,* http://entnemdept.ufl.edu/creatures/field/hornworm.htm (accessed September 27, 2016)
[9] The Top Worm, *All About Worms,* http://www.allaboutworms.com/tomato-and-tobacco-hornworms (accessed September 27, 2016)

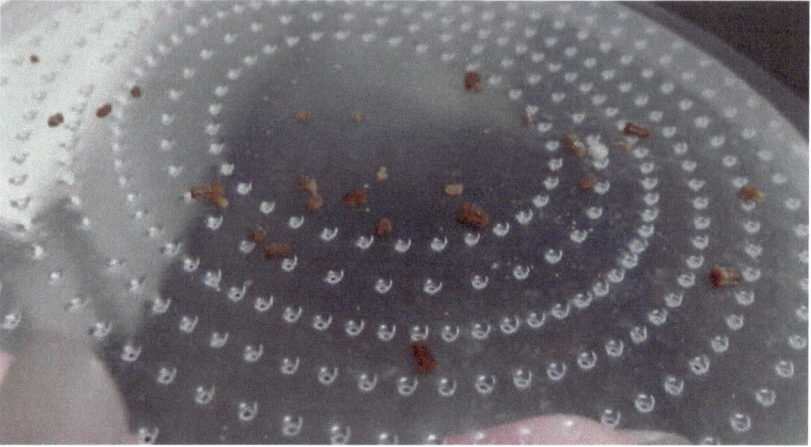

The hornworm may be large, but its poop is tiny. Here you can see a collection of it.

But don't take my word for it. Watch this video of a hornworm pooping.

https://youtu.be/lXUH6srlGTM

# MAMMALS AT THE ZOO

**Mammals** are a certain type of animal that—
1. Have fur or hair
2. Nurse their young
3. Give birth to live young

There are always exceptions, of course, but these are the main characteristics. Some examples of mammals you probably know are dogs, cats, bears, and tigers. Can you name any more?

This chapter brings you uncommon mammals, mammals you would probably only find in the zoo or a farm, or maybe even a circus. Believe it or not, you usually don't see the poop of the animals you go to visit at the zoo. Why is that? If it's a good zoo, then the caretakers remove the poop as often as several times a day.

**RED PANDA**

Everyone agrees that the red panda is an adorable creature, but what about its poop?

Before we examine that, let's look at its food. Why? Remember, you poop what you eat.

This is the food fed to the red pandas at Zoo Knoxville. It is made specifically to fulfill all of their nutritional requirements. After looking at their food, can you guess what their poop will look like?

It looks like two animals pooped in the same spot, doesn't it? The reason it looks that way is, again, because of what they're fed. It takes 4 hours for food to go through a red panda's system. Red pandas are **herbivores**, and so their poop can often be rainbow colored. When they eat pretty colored fruit (banana, watermelon, etc), they poop pretty colored logs.

Sometimes they have a really nasty mucus poop. That's because they have a mucus lining in their intestine that sheds periodically. It would be like farting boogers. So. Cool.

## SHREW

Have you seen poop like this before? Can you guess what animal made this poop? This is the poop of a shrew. The shrew is a small **rodent**, much like a gerbil or a hamster. The diets of these animals are very similar, so it's no wonder their poop is too. If you see this in your house, and you don't have a small rodent for a pet, chances are you have a mouse in the house. It's best to tell your parents what you saw, and *don't touch it!*

## NAKED MOLE RAT

Does this look like the poop from the shrew? It should. This is from the naked mole rat, another small rodent. However, the naked mole rat differs from other rodents in special ways.

Much like the social structure of bees, naked mole rat "families" have a Queen, workers, baby sitters, and other titled positions. Unlike the bees however, the job of the naked mole rat can change over time. They can get promoted.

They poop in a special way too. The whole family all poops in one latrine area. They then roll around in this latrine in order to all be of a uniform smell. That is their family's smell and how they recognize each other. Now just imagine your family washing in the toilet every morning so you all smelled the same. Do *NOT* try this at home!

## PORCUPINE

Can you guess what the porcupine eats by looking at its poop? While the porcupines at Zoo Knoxville chow down on nuts and berries, wild porcupines like to eat leaves and wood. What? Yes, wood. They've even been known to eat the paddles of unattended canoes.[10]

Want to see me petting a porcupine? Watch this—

https://youtu.be/amECOb37z2Y

---

[10] "Porcupines". National Geographic. www.nationalgeographic.com/animals/mammals/group/porcupines/. Accessed November 7, 2017.

**HEDGEHOG**

This poop was donated by Hoagie the hedgehog. While the hedgehog looks like a rodent, and therefore an herbivore, it is, in fact, an **insectivore**. I bet you can guess what an insectivore eats. Because its diet is so different from the other rodents, its poop looks different from theirs too[11].

---

[11] Sarah Glass (Curator), interviewed by Deborah Johnson, Knoxville, TN, April 2016

## BLACK BEAR

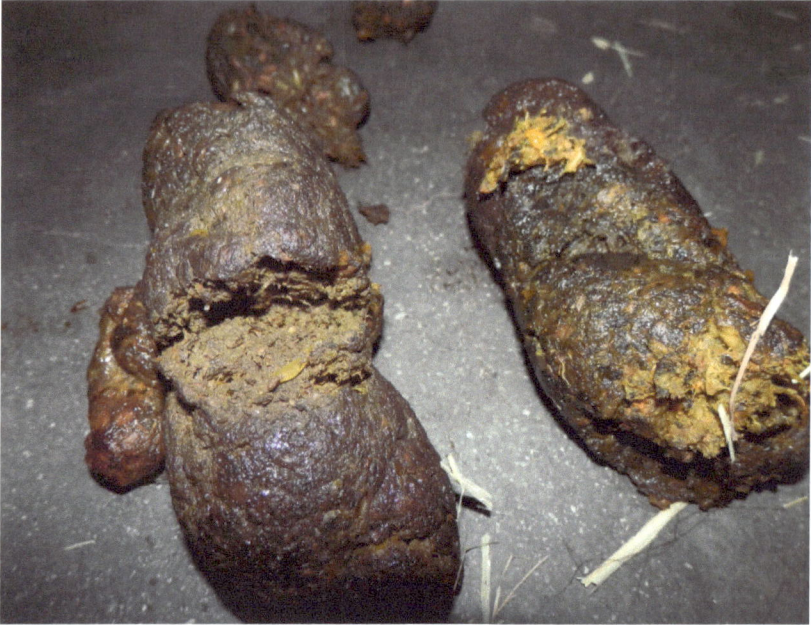

Picture taken by Rachel - common black bear

Black bears are **omnivores**. That means they eat *everything*. You may have noticed when camping all the warning signs and special garbage cans. That's to keep bears away from the garbage and you safe. If a bear gets used to finding its food in cars, garbage cans, or worse, tents, it will continue to look in those places for its food. That's very dangerous for you! Remember how your puppy jumps on you and begs for treats or food? Now, just imagine that bear sized!

# PIG

I know a pig sounds like a farm animal, but many pigs are kept as pets. Pigs are omnivores too and will eat just about anything they can find. Wild pigs are called wild boars, and they are highly dangerous. In fact, they are considered one of the most dangerous animals to hunt, even more dangerous than a bear.[12]

[12] McNamme, Brent. "The World's Most Dangerous Game to Hunt". November 9, 2011. *Hunter Safety Blog.* www.huntercourse.com/blog/2011/11/the-worlds-most-dangerous-game-to-hunt/. Accessed November 7, 2017.

**SKUNK**

   Skunks are **nocturnal** animals that used
to be classified with weasels, otters, and
badgers. However, they are now classified
into their own family simply because they
are just too stinky to fit in with the
others. Skunks eat a varied diet including
frogs, berries, and grubs.

# CHACOAN PECCARY

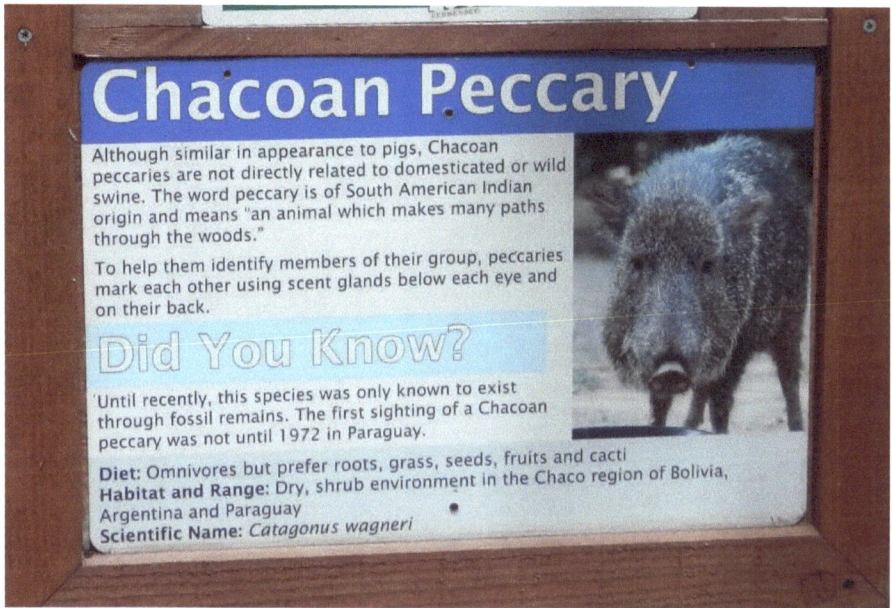

## Chacoan Peccary

Although similar in appearance to pigs, Chacoan peccaries are not directly related to domesticated or wild swine. The word peccary is of South American Indian origin and means "an animal which makes many paths through the woods."

To help them identify members of their group, peccaries mark each other using scent glands below each eye and on their back.

### Did You Know?

Until recently, this species was only known to exist through fossil remains. The first sighting of a Chacoan peccary was not until 1972 in Paraguay.

**Diet:** Omnivores but prefer roots, grass, seeds, fruits and cacti
**Habitat and Range:** Dry, shrub environment in the Chaco region of Bolivia, Argentina and Paraguay
**Scientific Name:** *Catagonus wagneri*

Peccaries can be found from the Southwest US all the way to South America. But the Chacoan peccary was once thought to be **extinct**. It was discovered alive in 1971 in Argentina and is now considered endangered. They mark their territory with smelly, milky stuff that might be confused with urate, but it's not. Think of it like a skunk smell instead of a type of waste product.

Chacoan peccaries poop in designated locations, much like the naked mole rat. Looks like a formula for chocolate chip cookies to me.

# CIVET CAT

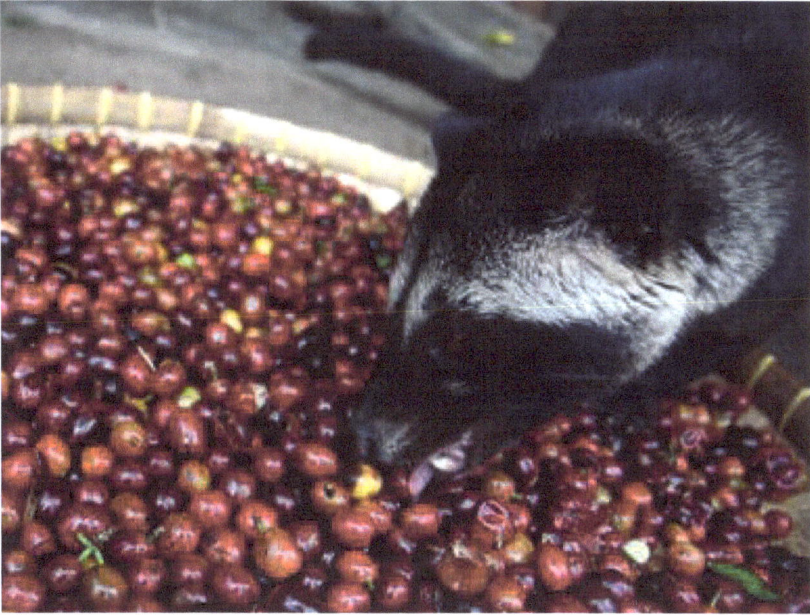

Photo by Ulet Ifansasti

The civet cat's poop is highly sought after, but not for reasons you would think. People like to drink this poop.

This is a picture of a civet cat eating coffee. Coffee beans are found inside these coffee "berries". If in Indonesia, these berries are often eaten by the civet cat. The cat is unable to digest the inside seed, what we call a coffee bean. So, after processing the fruity part of the berry, it poops out the beans, like this.

The beans are removed from the poop and washed before being roasted and served as coffee. This coffee is the most expensive in the world at $90 a cup![13] Wow! No thank you!

---

[13] Gloria Riviera, Nick Capote, Lauren Effron, *ABCNews*, http://abcnews.go.com/International/civet-cat-poop-coffee-worlds-expensive-brews-animal/story?id=30011989, (accessed September 27, 2016)

**ELEPHANT**

This is elephant poop. Chances are, if you've been to the zoo or a circus, you've seen elephant poop. But did you know that Zoo Knoxville removes 250 pounds of elephant poop per day?[14] That's more than twice what your whole body weighs!

[14] Amy Flew (Mammal Curator), interviewed by Deborah Johnson, Knoxville, TN, MONTH 2016

# PRAIRIE DOGS AND OTHER BURROWING ANIMALS

Prairie dog poop is extremely hard to find. Why? Because prairie dogs poop underground. Just like woodchucks, gophers, and groundhogs, prairie dogs have little bathrooms that they dig out in their burrows.

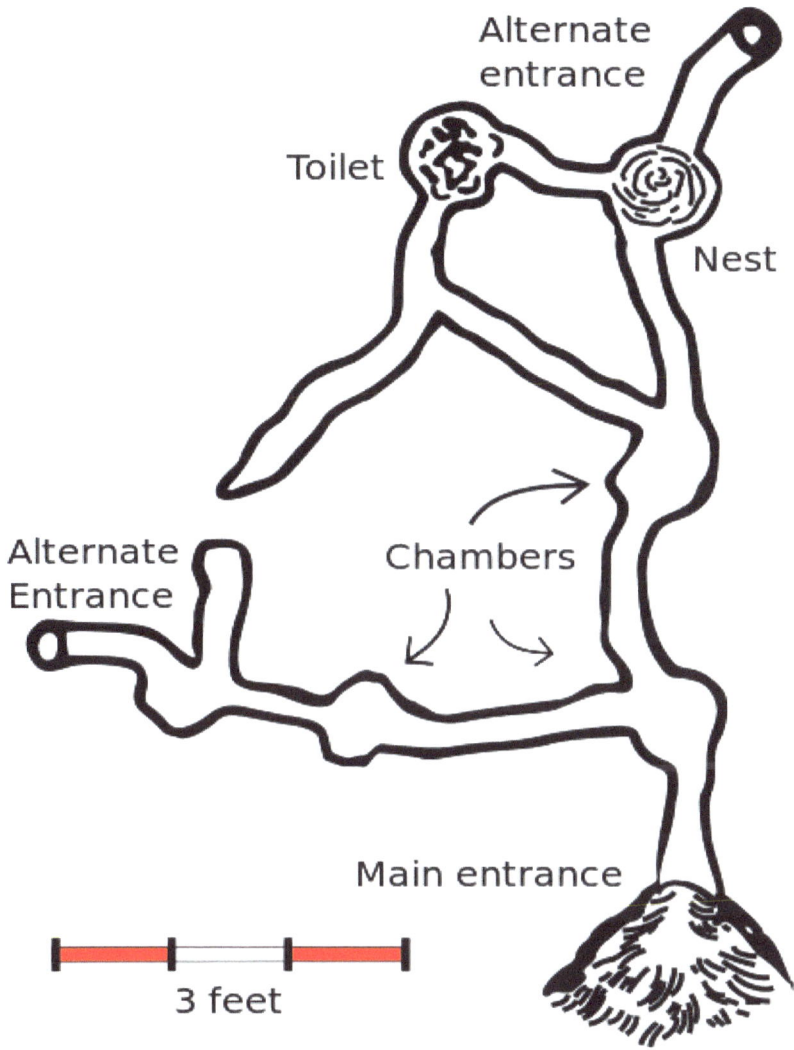

Alternate entrance

Toilet

Nest

Alternate Entrance

Chambers

Main entrance

3 feet

Photo courtesy of WikiCommons

Here is a diagram of a prairie dog's burrow. Do you see its toilet? It's close enough to the nest to be convenient, but far enough away to stay clean.

# RUMINANTS

**Ruminants** are a type of hooved mammal that "chews the cud". That's a really nice way of saying they eat their own vomit ... or worse.

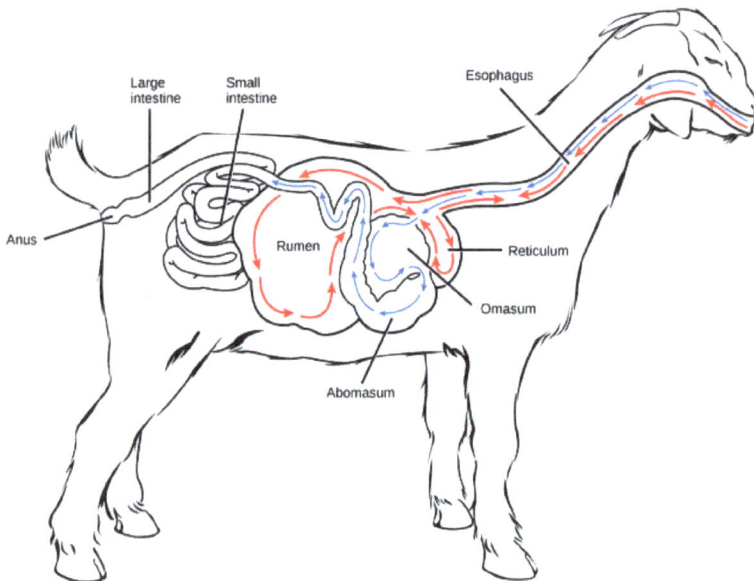

Photo from Lumen Learning

If you follow the arrows in the diagram above, you'll see that the food the sheep eats gets digested a little before it travels back into their mouth. They chew it some more and then swallow it again, where it finishes the process. Each one of the ruminants; sheep, goats, cows, deer, horses, etc regurgitate their food at a different point in their bodies. And guess what? All their poop looks very similar. To learn more about the digestion of ruminants, visit—
https://courses.lumenlearning.com/boundless-biology/chapter/digestive-systems/

**RABBIT**

Photo courtesy of The Rabbit House

This may look like poop, but it's not. It's called **cecotropes**, or "night feces", and it's from a rabbit. Late at night, when you're not watching, rabbits poop out their cecotropes and then eat them. Yes, you read that right.

Cecotropes are soft and full of nutrients that rabbits need to survive. If they don't eat them, they will actually die of **malnutrition**.[15]

Every animal is unique, and rabbits are specifically made to do this. If you tried to eat your own poop, you would get very very sick and could even die. So do **not** eat your own poop!!!

---

[15] Hess, Dr. Laurie, DVM, DABPV. *Why Does My Rabbit … Eat His Poop?*. VetStreet, May 7, 2015. http://www.vetstreet.com/our-pet-experts/why-does-my-rabbit-eat-his-poop (accessed November 9, 2017).

## HORSE

I was going to do a section on zebra poop, but it wasn't all that interesting. Zebra poop is basically horse poop, just more uniform. Why more uniform? Well, most horses live on a pasture. They eat grass in the warm months and hay in the cold. Sometimes, depending on their owner, they may eat grains. Wild zebras live in warmer climates with dry grasses for food. They don't get the moist, green grass that most American horses get.

The drier poop is, the harder it's going to be. Horse poop is wet enough to smoosh easily. But zebra poop is harder, and so it holds its shape better, making it appear more uniform than horse poop.

Now for the interesting part. Did you know that horse manure makes excellent heating fuel? Yes, horse poop, along with most other manures, can be dried and used for fuel in your wood stove or fireplace. Many people do this by shaping the manure into bricks, either with a mold or by hand, and then allowing it to dry. The drying time varies from hours to weeks, depending on the climate and the moisture content of the manure itself.

Photo courtesy of Takeaway of Wiki Commons

Here, hand sculpted water buffalo manure that has been thrown against a wall is allowed to dry. Once dried, these patties will be burned for cooking or heat.

Two excellent videos on utilizing manure for both fuel and construction can be found here—

fuel - https://youtu.be/VbLOdOYWUZU
construction -
https://youtu.be/rguawpCMzT4

In the US, many people make horse manure bricks to burn for heat in their wood stoves. One brick is said to last all night!

# GOAT

Goats and deer are very much alike. So is their poop. Remember why? That's right. They're both ruminants. Are you noticing a pattern with most of the ruminants? Remember how the rabbit and the horse also pooped in pellets?

# CAMEL

Camels do chew the cud, but they are called a **psuedoruminant.** "True" ruminants have stomachs with four chambers. Psuedoruminants, like camels and hippos, only have three chambers. Camel poop looks very similar to horse, and hippo poop is more like a cow or buffalo.

Thanks to the internet, hippopotamus poop has become well known, and here's why. Much like dogs and other animals, hippos like to mark their territory, but they have a more interesting way of doing it than your next-door neighbor's Chihuahua. They use their tails to fling their poop all over the place. Here's a video of one doing just that:

https://youtu.be/hy8nDb5nCak

## WATER BASED ANIMALS

The poop of marine life breaks down in the water too quickly to get good pictures normally, so this chapter is based on water *based* animals, not actual fish, etc. If you would like to know more about the poop of marine life, go to the end of this book where I have listed several links for various poops of marine life. In addition to the links at the end of this book, I recommend searching for crocodile, dolphin, and whale poop.

# PENGUIN

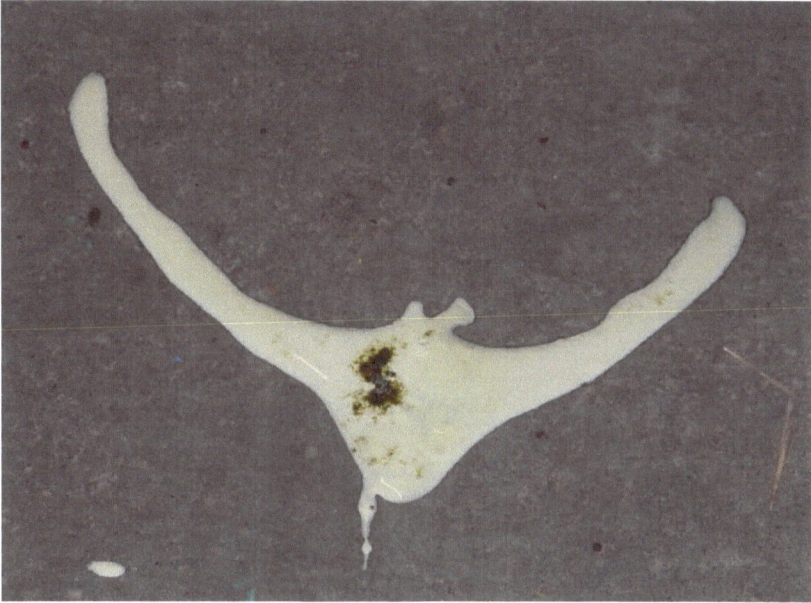

If I asked you what birds use to nest with, what would you say? Grass? Feathers? People's hair? What about poop? Yes, that V-shaped splatter of poop up there is nesting material for penguins.

A wild penguin's lifespan averages around 15 years, but they can live up to 50 in captivity. That's a long time for birds. However, despite such a long lifespan, penguins became endangered prior to artificial fertilizer. This was mainly due to man over harvesting their **guano**.

All guano makes amazing fertilizer, but penguin guano also makes amazing nesting material. It makes a squishy pad that keeps their eggs warm and snuggly. Remove that poopy pillow, and you have a hard, cold, rock. Eggs don't want to hatch on a hard, cold rock, not even penguin eggs. So, since they couldn't replace the

old, dying penguins with new, baby ones,
their population decreased.

# DUCK

This is poop from one of my own ducks. It's sitting on top of some chicken poop. Can you tell the difference? Unlike the solid chicken poop, duck poop is wet and sticky, like tar. Go back and look at the penguin poop. See how the duck's looks like a cross between the penguin and the chicken? Can you guess why?

That's right, diet. My ducks eat all the bugs and tasty things they can catch on their own from the outside, but that's usually not enough. I give all my birds free access to all the grain-based chicken feed they want so they never go hungry. Penguins eat mainly fish, not grain, and they stay in the water longer than ducks do on any given day. That's why their poop is so different in appearance and texture.

**OTTER**

Otters like their water clean, so they come out of the water to poop, like this one did.

**BEAVER**

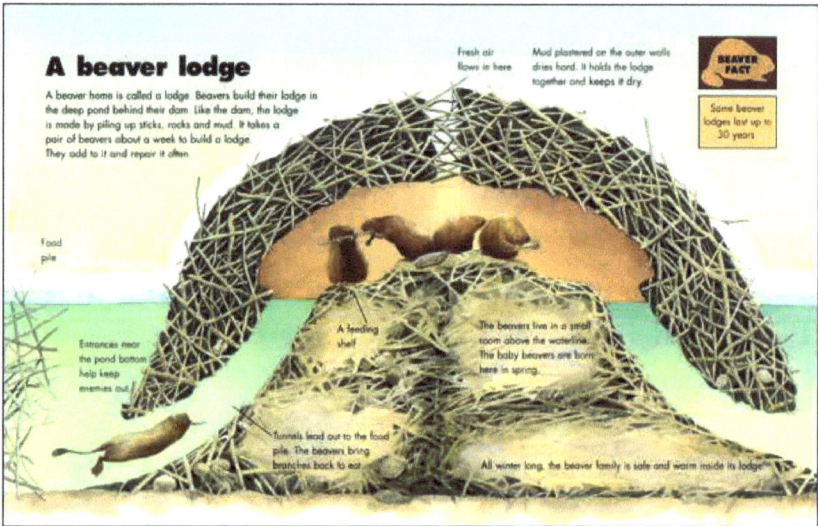

## A beaver lodge

A beaver home is called a lodge. Beavers build their lodge in the deep pond behind their dam. Like the dam, the lodge is made by piling up sticks, rocks and mud. It takes a pair of beavers about a week to build a lodge. They add to it and repair it often.

Fresh air flows in here

Mud plastered on the outer walls dries hard. It holds the lodge together and keeps it dry

**BEAVER FACT**

Some beaver lodges last up to 30 years

Food pile

Entrances near the pond bottom help keep enemies out.

A feeding shelf

The beavers live in a small room above the waterline. The baby beavers are born here in spring.

Tunnels lead out to the food pile. The beavers bring branches back to eat.

All winter long, the beaver family is safe and warm inside its lodge.

Diagram by Brendan McKee of WikiSpaces

Here is a diagram of a beaver's house, or "lodge". Do you see a toilet area anywhere? No, you don't. That's because beavers are clean animals and don't want their poop in their lodge, so they go outside to do their business.

As you can see, beaver poop floats. Beavers are extremely interesting animals.

That's all the poop for this book. I hope you have enjoyed learning about animal manure and are taking the next steps to learn more. The links to websites and videos are listed after the glossary, along with other resources to help you with your further exploration into the world of animal poop, including potential careers. Have fun!

If you liked this book, please support the author by leaving a review on your favorite review site.

For more works by N.A. Cauldron, visit http://nacauldron.com/

Thank you!

# GLOSSARY

**Arachnid–**any of a class of arthropods including the spiders, scorpions, mites, and ticks and having a segmented body divided into two regions of which the front part bears four pairs of legs but no antennae

**Carnivore-**a flesh-eating animal; an animal that eats meat

**Cecotropes-** caecal pellets, or night feces, are the product of the cecum, a part of the digestive system in mammals of the order lagomorpha, which includes two families: Leporidae (hares and rabbits), and Ochotonidae (pikas).

**Citrus-**any of a genus of often thorny trees and shrubs (as the orange, grapefruit, or lemon) grown in warm regions for their fruits; also : the fruit of a citrus

**Cloaca-**a chamber into which the intestinal, urinary, and reproductive canals empty in birds, reptiles, amphibians, and some fishes; also : a chamber like this in an invertebrate animal that serves the same purpose

**Digest-**to convert food into simpler forms that can be taken in and used by the body

**Endangered-**threatened with extinction <an endangered species>

**Excrete-**to separate and eliminate (waste) from the blood or tissues or from the active protoplasm usually in the form of sweat or urine

**Extinct-**no longer existing <an extinct species of animal>

**Feces-**bodily waste discharged through the anus; excrement

**Guano-**a substance composed chiefly of the excrement of seabirds or bats and used as a fertilizer

**Herbivore-**a plant-eating animal

**Hull-**the outer covering of a fruit or seed

**Insect-**any of a class of arthropods (as butterflies, true bugs, two-winged flies, bees, and grasshoppers) with the body clearly divided into a head, thorax, and abdomen, with three pairs of jointed legs, and usually with one or two pairs of wings

**Insectivore-**an insect-eating plant or animal

**Kidneys-**either of a pair of oval to bean-shaped organs located in the back part of the abdomen near the spine that give off waste products in the form of urine

**Mammal-**any of a class of warm-blooded vertebrates that include human beings and all other animals that nourish their young with milk produced by mammary glands and have the skin usually more or less covered with hair

**Malnutrition-**faulty nutrition especially due to inadequate intake of nutrients

**Monogamous-**marriage with only one person at a time

**Nocturnal-**active at night <nocturnal insects>

**Omnivore-**an animal that consumes both plant and animal matter for food

**Psuedoruminant-**a classification of animal based on its digestive tract differing from the ruminants. Hippopotamidae (comprising hippopotami) are ungulate mammals with a three-chambered stomach (ruminants have a four-chambered stomach)

**Quarantine-**a limiting or forbidding of movements of persons, goods, or animals that is designed to prevent the spread of disease or pests

**Regurgitate-**to throw or be thrown back or out again <regurgitate undigested food>

**Rodent-**any of an order of fairly small mammals (as mice, squirrels, or beavers) that have sharp front teeth used for gnawing

**Rumen-**the large first compartment of the stomach of a cud-chewing mammal (as a cow) in which cellulose is broken down by the action of microorganisms and in which food is stored prior to chewing

**Ruminant-**a cud-chewing mammal

**Scat-**the feces deposited by an animal \<bear scat\>

**Species-**a class of things of the same kind and with the same name kind, sort; a category of living things that ranks below a genus, is made up of related individuals able to produce fertile offspring, and is identified by a two-part scientific name

**Toxic-**of, relating to, or caused by a poison or toxin

**Toxin-**an antigenic poison or venom of plant or animal origin, especially one produced by or derived from microorganisms and causing disease when present at low concentration in the body.

**Urate-**a salt or ester of uric acid

**Urine-**waste material that is secreted by the kidneys, is rich in the end products of protein breakdown, and is usually a yellowish liquid in mammals but semisolid in birds and reptiles

# WEBSITES AND BOOKS

## JOBS IN STUDYING POOP
https://www.sciencenewsforstudents.org/article/cool-jobs-delving-dung

## JOBS IN WILDLIFE MANAGEMENT
http://www2.humboldt.edu/wildlife/wildcareers.html
http://learn.org/articles/What_are_Some_Popular_Careers_in_Wildlife_Management.html
http://study.com/articles/Wildlife_Environment_Manager_Job_Description_and_Info_for_Students_Considering_a_Career_in_Wildlife_Environment_Mgmt.html

## VIDEOS AND IMAGES OF OTHER POOPS
https://www.youtube.com/ has numerous videos on animals pooping. If there's an animal not mentioned in this book that you would like to see pooping, please visit YouTube. The internet can be a dangerous place. Please have your parents with you when you use it.
For specific videos and articles on marine life poop, here are some links:
Diagnosing the health of your pet fish based on its poop—
https://www.kokosgoldfish.com/GoldfishPoop.html
A video of a great white pooping
https://youtu.be/jfeHNy8Vxh4
Making octopus poop candy
https://youtu.be/uiMnmzVpPuY

## THE WOODLAND PARK ZOO FECAL FEST
https://www.zoo.org/conservation/fecalfest#.V6C2i-RWhaU

## PURCHASING OWL PELLETS
Some places will give you one for free, but if that's not possible, a simple online search for "buy owl pellet" will result in far more sites than I can list here. I suggest you find a supplier that fits your needs.

## JEFF THE NATURE GUY ON SCAT
https://youtu.be/QYrPczmHpq0

## ARACARI AND GOLDEN RETRIEVER
https://youtu.be/ZU3xc6q2QAA

## REGURGITATING AND OWL PELLET
https://youtu.be/PTLaxqmcYeY

## A SCREAMER CRESTED SCREAMER

https://youtu.be/SeaByXgH7nU

**A SNAKE POOPING**
https://youtu.be/LGm3tKOwWX4

**A GRASSHOPPER POOPING**
https://youtu.be/JsdEa_3L8GQ

**A HORNWORM POOPING**
https://youtu.be/lXUH6srlGTM

**PETTING A PORCUPINE**
https://youtu.be/amECOb37z2Y

**THE DIGESTION OF RUMINENTS**
https://courses.lumenlearning.com/boundless
-biology/chapter/digestive-systems/

**UTILIZING MANURE FOR FUEL**
https://youtu.be/VbLOdOYWUZU
**FOR CONSTRUCTION**
https://youtu.be/rguawpCMzT4

**HIPPOPOTAMUS POOPING**
https://youtu.be/hy8nDb5nCak

www.ingramcontent.com/pod-product-compliance
Lightning Source LLC
Chambersburg PA
CBHW040932030426
42336CB00001B/7